The Method of Archimedes

A Supplement to the Works of Archimedes

The Method of Archimedes

A Supplement to the Works of Archimedes

SIR THOMAS HEATH

NEW YORK

The Method of Archimedes: A Supplement to the Works of Archimedes
Cover Copyright © 2007 by Cosimo, Inc.

The Method of Archimedes: A Supplement to the Works of Archimedes was originally published in 1897.

For information, address:
P.O. Box 416, Old Chelsea Station
New York, NY 10011

or visit our website at:
www.cosimobooks.com

Ordering Information:
Cosimo publications are available at online bookstores. They may also be purchased for educational, business or promotional use:
- *Bulk orders:* special discounts are available on bulk orders for reading groups, organizations, businesses, and others. For details contact Cosimo Special Sales at the address above or at info@cosimobooks.com.
- *Custom-label orders:* we can prepare selected books with your cover or logo of choice. For more information, please contact Cosimo at info@cosimobooks.com.

Cover Design by www.popshopstudio.com

ISBN: 978-1-60206-391-4

INTRODUCTORY NOTE

From the point of view of the student of Greek mathematics there has been, in recent years, no event comparable in interest with the discovery by Heiberg in 1906 of a Greek MS. containing, among other works of Archimedes, substantially the whole of a treatise which was formerly thought to be irretrievably lost. The full description of the MS. as given in the preface to Vol. I. (1910) of the new edition of Heiberg's text of Archimedes now in course of publication is—

Codex rescriptus Metochii Constantinopolitani S. Sepulchri monasterii Hierosolymitani 355, 4to.

Heiberg has told the story of his discovery of this MS. and given a full description of it*. His attention having been called to a notice in Vol. IV. (1899) of the Ἱεροσολυμιτικὴ Βιβλιοθήκη of Papadopulos Kerameus relating to a palimpsest of mathematical content, he at once inferred from a few specimen lines which were quoted that the MS. must contain something by Archimedes. As the result of inspection, at Constantinople, of the MS. itself, and by means of a photograph taken of it, he was able to see what it contained and to decipher much of the contents. This was in the year 1906, and he inspected the MS. once more in 1908. With the exception of the last leaves, 178 to 185, which are of paper of the 16th century, the MS. is of parchment and contains writings of Archimedes copied in a good hand of the 10th century, in two columns. An attempt was made (fortunately with only partial success) to wash out the old writing, and then the parchment was used again, for the purpose of writing a Euchologion thereon, in the 12th—13th or 13th—14th centuries. The earlier writing appears with more or less clearness on most of the 177 leaves; only 29 leaves are destitute of any trace of such writing; from 9 more it was hopelessly washed off; on a few more leaves only a few words can be made out; and again some 14 leaves have old writing

* *Hermes* XLII. 1907, pp. 235 sq.

upon them in a different hand and with no division into columns. All the rest is tolerably legible with the aid of a magnifying glass. Of the treatises of Archimedes which are found in other MSS., the new MS. contains, in great part, the books *On the Sphere and Cylinder*, almost the whole of the work *On Spirals*, and some parts of the *Measurement of a Circle* and of the books *On the Equilibrium of Planes*. But the important fact is that it contains (1) a considerable proportion of the work *On Floating Bodies* which was formerly supposed to be lost so far as the Greek text is concerned and only to have survived in the translation by Wilhelm von Mörbeke, and (2), most precious of all, the greater part of the book called, according to its own heading, Ἔφοδος and elsewhere, alternatively, Ἐφόδιον or Ἐφοδικόν, meaning *Method*. The portion of this latter work contained in the MS. has already been published by Heiberg (1) in Greek* and (2) in a German translation with commentary by Zeuthen†. The treatise was formerly only known by an allusion to it in Suidas, who says that Theodosius wrote a commentary upon it; but the *Metrica* of Heron, newly discovered by R. Schöne and published in 1903, quotes three propositions from it‡, including the two main propositions enunciated by Archimedes at the beginning as theorems novel in character which the method furnished a means of investigating. Lastly the MS. contains two short propositions, in addition to the preface, of a work called *Stomachion* (as it might be "Neck-Spiel" or "Quäl-Geist") which treated of a sort of Chinese puzzle known afterwards by the name of "loculus Archimedius"; it thus turns out that this puzzle, which Heiberg was formerly disinclined to attribute to Archimedes§, is really genuine.

The *Method*, so happily recovered, is of the greatest interest for the following reason. Nothing is more characteristic of the classical works of the great geometers of Greece, or more tantalising, than the absence of any indication of the steps by which they worked their way to the discovery of their great theorems. As they have come down to us, these theorems are finished masterpieces which leave no traces of any rough-hewn stage, no hint of the method by which they were evolved. We cannot but suppose that the

* *Hermes* XLII. 1907, pp. 243—297.
† *Bibliotheca Mathematica* VII₃, 1906-7, pp. 321—363.
‡ *Heronis Alexandrini opera*, Vol. III. 1903, pp. 80, 17; 130, 15; 130, 25.
§ Vide *The Works of Archimedes*, p. xxii.

Greeks had some method or methods of analysis hardly less powerful than those of modern analysis; yet, in general, they seem to have taken pains to clear away all traces of the machinery used and all the litter, so to speak, resulting from tentative efforts, before they permitted themselves to publish, in sequence carefully thought out, and with definitive and rigorously scientific proofs, the results obtained. A partial exception is now furnished by the *Method*; for here we have a sort of lifting of the veil, a glimpse of the interior of Archimedes' workshop as it were. He tells us how he discovered certain theorems in quadrature and cubature, and he is at the same time careful to insist on the difference between (1) the means which may be sufficient to suggest the truth of theorems, although not furnishing scientific proofs of them, and (2) the rigorous demonstrations of them by irrefragable geometrical methods which must follow before they can be finally accepted as established; to use Archimedes' own terms, the former enable theorems to be *investigated* (θεωρεῖν) but not to be *proved* (ἀποδεικνύναι). The mechanical method, then, used in our treatise and shown to be so useful for the discovery of theorems is distinctly said to be incapable of furnishing proofs of them; and Archimedes promises to add, as regards the two main theorems enunciated at the beginning, the necessary supplement in the shape of the formal geometrical proof. One of the two geometrical proofs is lost, but fragments of the other are contained in the MS. which are sufficient to show that the method was the orthodox method of exhaustion in the form in which Archimedes applies it elsewhere, and to enable the proof to be reconstructed.

The rest of this note will be best understood after the treatise itself has been read; but the essential features of the mechanical method employed by Archimedes are these. Suppose X to be a plane or solid figure, the area or content of which has to be found. The method is to weigh infinitesimal elements of X (with or without the addition of the corresponding elements of another figure C) against the corresponding elements of a figure B, B and C being such figures that their areas or volumes, and the position of the centre of gravity of B, are known beforehand. For this purpose the figures are first placed in such a position that they have, as common diameter or axis, one and the same straight line; if then the infinitesimal elements are sections of the figures made by parallel planes perpendicular (in general) to the axis and cutting the figures, the centres of gravity of all the elements lie at one point or other on the common diameter or axis. This diameter or axis is produced and is imagined to be the bar or lever of a balance. It is sufficient to take the simple case where the elements of X alone are weighed against the elements of another figure B. The elements which correspond to one another are the sections of X and B respectively by any one plane perpendicular (in general) to the diameter or axis and cutting both figures; the elements are spoken of as straight lines in the case of plane figures and as plane areas in the case of solid figures. Although Archimedes calls the elements straight lines and plane areas respectively, they are of course, in the first case, indefinitely narrow strips (areas) and, in the second case, indefinitely thin plane laminae (solids); but the breadth or thickness (dx, as we might call it) does not enter into the calculation because it is regarded as the same in each of the two corresponding elements which are separately weighed against each other, and therefore divides out. The number of the elements in each figure is infinite, but Archimedes has no need to say this; he merely says that X and B are *made up* of *all* the elements in them respectively, i.e. of the straight lines in the case of areas and of the plane areas in the case of solids.

The object of Archimedes is so to arrange the balancing of the elements that the elements of X are all applied at *one point* of the lever, while the elements of B operate at different points, namely where they actually are in the first instance. He contrives therefore to move the elements of X away from their first position and to concentrate them at one point on the lever, while the elements of B are left where they are, and so operate at their respective centres of gravity. Since the centre of gravity of B as a whole is known, as well as its area or volume, it may then be supposed to act as one mass applied at its centre of gravity; and consequently, taking the whole bodies X and B as ultimately placed respectively, we know the distances of the two centres of gravity from the fulcrum or point of suspension of the lever, and also the area or volume of B. Hence the area or volume of X is found. The method may be applied, conversely, to the problem of finding the centre of gravity of X when its area or volume is known beforehand; in this case it is necessary that the elements of X, and therefore X itself, should be weighed *in the places where they are*, and that the figures the elements of which are moved to one single point of the lever, to be weighed there, should be other figures and not X.

The method will be seen to be, not *integration*, as certain geometrical proofs in the great treatises actually are, but a clever device for *avoiding* the particular integration which would naturally be used to find directly the area or volume required, and making the solution depend, instead, upon *another* integration the result of which is already known. Archimedes deals with *moments* about the point of suspension of the lever, i.e. the products of the elements of area or volume into the distances between the point of suspension of the lever and the centres of gravity of the elements respectively; and, as we said above, while these distances are different for all the elements of B, he contrives, by moving the elements of X, to make them the same for all the elements of X in their final position. He assumes, as known, the fact that the sum of the moments of each particle of the figure B acting at the point where it is placed is equal to the moment of the whole figure applied as one mass at one point, its centre of gravity.

Suppose now that the element of X is $u\,.\,dx$, u being the length or area of a section of X by one of a whole series of parallel planes cutting the lever at right angles, x being measured along the lever (which is the common axis of the two figures) from the point of suspension of the lever as origin. This element is then supposed to be placed on the lever at a constant distance, say a, from the origin and on the opposite side of it from B. If $u'\,.\,dx$ is the corresponding element of B cut off by the same plane and x its distance from the origin, Archimedes' argument establishes the equation

$$a \int_h^k u\,dx = \int_h^k x u'\,dx.$$

Now the second integral is known because the area or volume of the figure B (say a triangle, a pyramid, a prism, a sphere, a cone, or a cylinder) is known, and it can be supposed to be applied as one mass at its centre of gravity, which is also known; the integral is equal to bU, where b is the distance of the centre of gravity from the point of suspension of the lever, and U is the area or content of B. Hence

$$\text{the area or volume of } X = \frac{bU}{a}.$$

In the case where the elements of X are weighed along with the corresponding elements of another figure C against corresponding elements of B, we have, if v be the element of C, and V its area or content,

$$a \int_h^k u\,dx + a \int_h^k v\,dx = \int_h^k x\,u'\,dx$$

and (area or volume of $X + V$) $a = bU$.

In the particular problems dealt with in the treatise h is always $= 0$, and k is often, but not always, equal to a.

Our admiration of the genius of the greatest mathematician of antiquity must surely be increased, if that were possible, by a perusal of the work before us. Mathematicians will doubtless agree that it is astounding that Archimedes, writing (say) about 250 B.C., should have been able to solve such problems as those of finding the volume and the centre of gravity of any segment of a sphere, and the centre of gravity of a semicircle, by a method so simple, a method too (be it observed) which would be quite rigorous enough for us to-day, although it did not satisfy Archimedes himself.

Apart from the mathematical content of the book, it is interesting, not only for Archimedes' explanations of the course which his investigations took, but also for the allusion to *Democritus* as the discoverer of the theorem that the volumes of a pyramid and a cone are one-third of the volumes of a prism and a cylinder respectively which have the same base and equal height. These propositions had always been supposed to be due to Eudoxus, and indeed Archimedes himself has a statement to this effect[*]. It now appears that, though Eudoxus was the first to prove them scientifically, Democritus was the first to assert their truth. I have elsewhere[†] made a suggestion as to the probable course of Democritus' argument, which, on Archimedes' view, did not amount to a proof of the propositions; but it may be well to re-state it here. Plutarch, in a well-known passage[‡], speaks of Democritus as having raised the following question in natural philosophy (φυσικῶς): "if a cone were cut by a plane parallel to the base [by which is clearly meant a plane indefinitely near to the base], what must we think of the surfaces of the sections? Are they equal or unequal? For, if they are unequal, they will make the cone irregular, as having many indentations, like steps, and unevennesses; but, if they are equal, the sections will be equal, and the cone will appear to have the property of the cylinder and to be made up of equal, not unequal,

---

[*] *On the Sphere and Cylinder*, Preface to Book I.
[†] *The Thirteen Books of Euclid's Elements*, Vol. III. p. 368.
[‡] Plutarch, *De Comm. Not. adv. Stoicos* XXXIX. 3.

circles, which is very absurd." The phrase "made up of equal... circles" (ἐξ ἴσων συγκείμενος...κύκλων) shows that Democritus already had the idea of a solid being the sum of an infinite number of parallel planes, or indefinitely thin laminae, indefinitely near together: a most important anticipation of the same thought which led to such fruitful results in Archimedes. If then we may make a conjecture as to Democritus' argument with regard to a pyramid, it seems probable that he would notice that, if two pyramids of the same height and with equal triangular bases are respectively cut by planes parallel to the base and dividing the heights in the same ratio, the corresponding sections of the two pyramids are equal, whence he would infer that the pyramids are equal because they are the sums of the same infinite numbers of equal plane sections or indefinitely thin laminae. (This would be a particular anticipation of Cavalieri's proposition that the areal or solid contents of two figures are equal if two sections of them taken at the same height, whatever the height may be, always give equal straight lines or equal surfaces respectively.) And Democritus would of course see that the three pyramids into which a prism on the same base and of equal height with the original pyramid is divided (as in Eucl. XII. 7) satisfy, in pairs, this test of equality, so that the pyramid would be one third part of the prism. The extension to a pyramid with a polygonal base would be easy. And Democritus may have stated the proposition for the cone (of course without an absolute proof) as a natural inference from the result of increasing indefinitely the number of sides in a regular polygon forming the base of a pyramid.

In accordance with the plan adopted in *The Works of Archimedes*, I have marked by inverted commas the passages which, on account of their importance, historically or otherwise, I have translated literally from the Greek; the rest of the tract is reproduced in modern notation and phraseology. Words and sentences in square brackets represent for the most part Heiberg's conjectural restoration (in his German translation) of what may be supposed to have been written in the places where the MS. is illegible; in a few cases where the gap is considerable a note in brackets indicates what the missing passage presumably contained and, so far as necessary, how the deficiency may be made good.

T. L. H.

7 *June* 1912.

THE METHOD OF ARCHIMEDES TREATING OF MECHANICAL PROBLEMS— TO ERATOSTHENES

"Archimedes to Eratosthenes greeting.

I sent you on a former occasion some of the theorems discovered by me, merely writing out the enunciations and inviting you to discover the proofs, which at the moment I did not give. The enunciations of the theorems which I sent were as follows.

1. If in a right prism with a parallelogrammic base a cylinder be inscribed which has its bases in the opposite parallelograms*, and its sides [i.e. four generators] on the remaining planes (faces) of the prism, and if through the centre of the circle which is the base of the cylinder and (through) one side of the square in the plane opposite to it a plane be drawn, the plane so drawn will cut off from the cylinder a segment which is bounded by two planes and the surface of the cylinder, one of the two planes being the plane which has been drawn and the other the plane in which the base of the cylinder is, and the surface being that which is between the said planes; and the segment cut off from the cylinder is one sixth part of the whole prism.

2. If in a cube a cylinder be inscribed which has its bases in the opposite parallelograms† and touches with its surface the remaining four planes (faces), and if there also be inscribed in the same cube another cylinder which has its bases in other parallelograms and touches with its surface the remaining four planes (faces), then the figure bounded by the surfaces of the cylinders, which is within both cylinders, is two-thirds of the whole cube.

Now these theorems differ in character from those communicated before; for we compared the figures then in question,

---

* The parallelograms are apparently *squares*. † i.e. squares.

conoids and spheroids and segments of them, in respect of size, with figures of cones and cylinders: but none of those figures have yet been found to be equal to a solid figure bounded by planes; whereas each of the present figures bounded by two planes and surfaces of cylinders is found to be equal to one of the solid figures which are bounded by planes. The proofs then of these theorems I have written in this book and now send to you. Seeing moreover in you, as I say, an earnest student, a man of considerable eminence in philosophy, and an admirer [of mathematical inquiry], I thought fit to write out for you and explain in detail in the same book the peculiarity of a certain method, by which it will be possible for you to get a start to enable you to investigate some of the problems in mathematics by means of mechanics. This procedure is, I am persuaded, no less useful even for the proof of the theorems themselves; for certain things first became clear to me by a mechanical method, although they had to be demonstrated by geometry afterwards because their investigation by the said method did not furnish an actual demonstration. But it is of course easier, when we have previously acquired, by the method, some knowledge of the questions, to supply the proof than it is to find it without any previous knowledge. This is a reason why, in the case of the theorems the proof of which Eudoxus was the first to discover, namely that the cone is a third part of the cylinder, and the pyramid of the prism, having the same base and equal height, we should give no small share of the credit to Democritus who was the first to make the assertion with regard to the said figure* though he did not prove it. I am myself in the position of having first made the discovery of the theorem now to be published [by the method indicated], and I deem it necessary to expound the method partly because I have already spoken of it† and I do not want to be thought to have uttered vain words, but

---

* περὶ τοῦ εἰρημένου σχήματος, in the singular. Possibly Archimedes may have thought of the case of the pyramid as being the more fundamental and as really involving that of the cone. Or perhaps "figure" may be intended for "type of figure."

† Cf. Preface to *Quadrature of Parabola*.

equally because I am persuaded that it will be of no little service to mathematics; for I apprehend that some, either of my contemporaries or of my successors, will, by means of the method when once established, be able to discover other theorems in addition, which have not yet occurred to me.

First then I will set out the very first theorem which became known to me by means of mechanics, namely that

Any segment of a section of a right-angled cone (i.e. a parabola) is four-thirds of the triangle which has the same base and equal height,

and after this I will give each of the other theorems investigated by the same method. Then, at the end of the book, I will give the geometrical [proofs of the propositions]...

[I premise the following propositions which I shall use in the course of the work.]

1. If from [one magnitude another magnitude be subtracted which has not the same centre of gravity, the centre of gravity of the remainder is found by] producing [the straight line joining the centres of gravity of the whole magnitude and of the subtracted part in the direction of the centre of gravity of the whole] and cutting off from it a length which has to the distance between the said centres of gravity the ratio which the weight of the subtracted magnitude has to the weight of the remainder.

[*On the Equilibrium of Planes*, I. 8]

2. If the centres of gravity of any number of magnitudes whatever be on the same straight line, the centre of gravity of the magnitude made up of all of them will be on the same straight line. [Cf. *Ibid.* I. 5]

3. The centre of gravity of any straight line is the point of bisection of the straight line. [Cf. *Ibid.* I. 4]

4. The centre of gravity of any triangle is the point in which the straight lines drawn from the angular points of the triangle to the middle points of the (opposite) sides cut one another. [*Ibid.* I. 13, 14]

5. The centre of gravity of any parallelogram is the point in which the diagonals meet. [*Ibid.* I. 10]

6. The centre of gravity of a circle is the point which is also the centre [of the circle].

7. The centre of gravity of any cylinder is the point of bisection of the axis.

8. The centre of gravity of any cone is [the point which divides its axis so that] the portion [adjacent to the vertex is] triple [of the portion adjacent to the base].

[All these propositions have already been] proved*. [Besides these I require also the following proposition, which is easily proved:

If in two series of magnitudes those of the first series are, in order, proportional to those of the second series and further] the magnitudes [of the first series], either all or some of them, are in any ratio whatever [to those of a third series], and if the magnitudes of the second series are in the same ratio to the corresponding magnitudes [of a fourth series], then the sum of the magnitudes of the first series has to the sum of the selected magnitudes of the third series the same ratio which the sum of the magnitudes of the second series has to the sum of the (correspondingly) selected magnitudes of the fourth series. [*On Conoids and Spheroids*, Prop. 1.]"

Proposition 1.

Let ABC be a segment of a parabola bounded by the straight line AC and the parabola ABC, and let D be the middle point of AC. Draw the straight line DBE parallel to the axis of the parabola and join AB, BC.

Then shall the segment ABC be $\tfrac{4}{3}$ of the triangle ABC.

From A draw AKF parallel to DE, and let the tangent to the parabola at C meet DBE in E and AKF in F. Produce CB to meet AF in K, and again produce CK to H, making KH equal to CK.

---

* The problem of finding the centre of gravity of a cone is not solved in any extant work of Archimedes. It may have been solved either in a separate treatise, such as the περὶ ζυγῶν, which is lost, or perhaps in a larger mechanical work of which the extant books *On the Equilibrium of Planes* formed only a part.

Consider CH as the bar of a balance, K being its middle point.

Let MO be any straight line parallel to ED, and let it meet CF, CK, AC in M, N, O and the curve in P.

Now, since CE is a tangent to the parabola and CD the semi-ordinate,
$$EB = BD;$$
"for this is proved in the Elements [of Conics]*."

Since FA, MO are parallel to ED, it follows that
$$FK = KA, \quad MN = NO.$$
Now, by the property of the parabola, "proved in a lemma,"
$MO : OP = CA : AO$ [Cf. *Quadrature of Parabola*, Prop. 5]
$\qquad\qquad\; = CK : KN$ [Eucl. VI. 2]
$\qquad\qquad\; = HK : KN.$

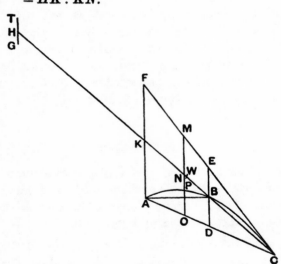

Take a straight line TG equal to OP, and place it with its centre of gravity at H, so that $TH = HG$; then, since N is the centre of gravity of the straight line MO,

and $\qquad\qquad MO : TG = HK : KN,$

---

* i.e. the works on conics by Aristaeus and Euclid. Cf. the similar expression in *On Conoids and Spheroids*, Prop. 3, and *Quadrature of Parabola*, Prop. 3.

it follows that TG at H and MO at N will be in equilibrium about K.   [*On the Equilibrium of Planes*, I. 6, 7]

Similarly, for all other straight lines parallel to DE and meeting the arc of the parabola, (1) the portion intercepted between FC, AC with its middle point on KC and (2) a length equal to the intercept between the curve and AC placed with its centre of gravity at H will be in equilibrium about K.

Therefore K is the centre of gravity of the whole system consisting (1) of all the straight lines as MO intercepted between FC, AC and placed as they actually are in the figure and (2) of all the straight lines placed at H equal to the straight lines as PO intercepted between the curve and AC.

And, since the triangle CFA is made up of all the parallel lines like MO,

and the segment CBA is made up of all the straight lines like PO within the curve,

it follows that the triangle, placed where it is in the figure, is in equilibrium about K with the segment CBA placed with its centre of gravity at H.

Divide KC at W so that $CK = 3KW$;

then W is the centre of gravity of the triangle ACF; "for this is proved in the books on equilibrium" (ἐν τοῖς ἰσορροπικοῖς).
                [Cf. *On the Equilibrium of Planes* I. 15]

Therefore $\triangle ACF$: (segment ABC) $= HK : KW$

$= 3 : 1$.

Therefore   segment $ABC = \frac{1}{3} \triangle ACF$.

But $\triangle ACF = 4 \triangle ABC$.

Therefore   segment $ABC = \frac{4}{3} \triangle ABC$.

"Now the fact here stated is not actually demonstrated by the argument used; but that argument has given a sort of indication that the conclusion is true. Seeing then that the theorem is not demonstrated, but at the same time suspecting that the conclusion is true, we shall have recourse to the geometrical demonstration which I myself discovered and have already published*."

Proposition 2.

We can investigate by the same method the propositions that

(1) *Any sphere is (in respect of solid content) four times the cone with base equal to a great circle of the sphere and height equal to its radius; and*

(2) *the cylinder with base equal to a great circle of the sphere and height equal to the diameter is $1\frac{1}{2}$ times the sphere.*

(1) Let $ABCD$ be a great circle of a sphere, and AC, BD diameters at right angles to one another.

Let a circle be drawn about BD as diameter and in a plane perpendicular to AC, and on this circle as base let a cone be described with A as vertex. Let the surface of this cone be produced and then cut by a plane through C parallel to its base; the section will be a circle on EF as diameter. On this circle as base let a cylinder be erected with height and axis AC, and produce CA to H, making AH equal to CA.

Let CH be regarded as the bar of a balance, A being its middle point.

Draw any straight line MN in the plane of the circle $ABCD$ and parallel to BD. Let MN meet the circle in O, P, the diameter AC in S, and the straight lines AE, AF in Q, R respectively. Join AO.

* The word governing τὴν γεωμετρουμένην ἀπόδειξιν in the Greek text is τάξομεν, a reading which seems to be doubtful and is certainly difficult to translate. Heiberg translates as if τάξομεν meant "we shall give lower down" or "later on," but I agree with Th. Reinach (*Revue générale des sciences pures et appliquées*, 30 November 1907, p. 918) that it is questionable whether Archimedes would really have written out in full once more, as an appendix, a proof which, as he says, had already been published (i.e. presumably in the *Quadrature of a Parabola*). τάξομεν, if correct, should apparently mean "we shall appoint," "prescribe" or "assign."

Through MN draw a plane at right angles to AC; this plane will cut the cylinder in a circle with diameter MN, the sphere in a circle with diameter OP, and the cone in a circle with diameter QR.

Now, since $MS = AC$, and $QS = AS$,
$$MS \cdot SQ = CA \cdot AS$$
$$= AO^2$$
$$= OS^2 + SQ^2.$$

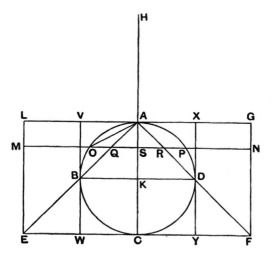

And, since $HA = AC$,
$$HA : AS = CA : AS$$
$$= MS : SQ$$
$$= MS^2 : MS \cdot SQ$$
$$= MS^2 : (OS^2 + SQ^2), \text{ from above,}$$
$$= MN^2 : (OP^2 + QR^2)$$
$$= (\text{circle, diam. } MN) : (\text{circle, diam. } OP$$
$$+ \text{circle, diam. } QR).$$

That is,

$HA : AS = (\text{circle in cylinder}) : (\text{circle in sphere} + \text{circle in cone}).$

Therefore the circle in the cylinder, placed where it is, is in equilibrium, about A, with the circle in the sphere together with the circle in the cone, if both the latter circles are placed with their centres of gravity at H.

Similarly for the three corresponding sections made by a plane perpendicular to AC and passing through any other straight line in the parallelogram LF parallel to EF.

If we deal in the same way with all the sets of three circles in which planes perpendicular to AC cut the cylinder, the sphere and the cone, and which make up those solids respectively, it follows that the cylinder, in the place where it is, will be in equilibrium about A with the sphere and the cone together, when both are placed with their centres of gravity at H.

Therefore, since K is the centre of gravity of the cylinder,

$$HA : AK = (\text{cylinder}) : (\text{sphere} + \text{cone } AEF).$$

But $HA = 2AK$;

therefore $\quad$ cylinder $= 2$ (sphere + cone AEF).

Now $\quad$ cylinder $= 3$ (cone AEF); $\qquad$ [Eucl. XII. 10]

therefore $\quad$ cone $AEF = 2$ (sphere).

But, since $EF = 2BD$,

cone $AEF = 8$ (cone ABD);

therefore $\quad$ sphere $= 4$ (cone ABD).

(2) Through B, D draw VBW, XDY parallel to AC; and imagine a cylinder which has AC for axis and the circles on VX, WY as diameters for bases.

Then $\quad$ cylinder $VY = 2$ (cylinder VD)

$= 6$ (cone ABD) $\qquad$ [Eucl. XII. 10]

$= \frac{3}{2}$ (sphere), from above.

Q.E.D.

"From this theorem, to the effect that a sphere is four times as great as the cone with a great circle of the sphere as base and with height equal to the radius of the sphere, I conceived the notion that the surface of any sphere is four times as great as a great circle in it; for, judging from the fact that any circle is equal to a triangle with base equal to the circumference and height equal to the radius of the circle, I apprehended that, in like manner, any sphere is equal to a cone with base equal to the surface of the sphere and height equal to the radius*."

Proposition 3.

By this method we can also investigate the theorem that

A cylinder with base equal to the greatest circle in a spheroid and height equal to the axis of the spheroid is $1\frac{1}{2}$ times the spheroid;

and, when this is established, it is plain that

If any spheroid be cut by a plane through the centre and at right angles to the axis, the half of the spheroid is double of the cone which has the same base and the same axis as the segment (i.e. *the half of the spheroid*).

Let a plane through the axis of a spheroid cut its surface in the ellipse $ABCD$, the diameters (i.e. axes) of which are AC, BD; and let K be the centre.

Draw a circle about BD as diameter and in a plane perpendicular to AC;

imagine a cone with this circle as base and A as vertex produced and cut by a plane through C parallel to its base; the section will be a circle in a plane at right angles to AC and about EF as diameter.

Imagine a cylinder with the latter circle as base and axis AC; produce CA to H, making AH equal to CA.

Let HC be regarded as the bar of a balance, A being its middle point.

In the parallelogram LF draw any straight line MN parallel to EF meeting the ellipse in O, P and AE, AF, AC in Q, R, S respectively.

---

* That is to say, Archimedes originally solved the problem of finding the solid content of a sphere before that of finding its surface, and he inferred the result of the latter problem from that of the former. Yet in *On the Sphere and Cylinder* I. the surface is independently found (Prop. 33) and *before* the volume, which is found in Prop. 34: another illustration of the fact that the order of propositions in the treatises of the Greek geometers as finally elaborated does not necessarily follow the order of discovery.

If now a plane be drawn through MN at right angles to AC, it will cut the cylinder in a circle with diameter MN, the spheroid in a circle with diameter OP, and the cone in a circle with diameter QR.

Since $HA = AC$,
$$HA : AS = CA : AS$$
$$= EA : AQ$$
$$= MS : SQ.$$

Therefore $\quad HA : AS = MS^2 : MS \cdot SQ.$

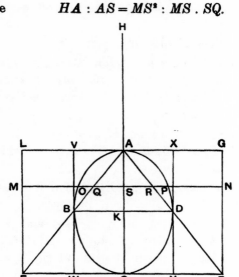

But, by the property of the ellipse,
$$AS \cdot SC : SO^2 = AK^2 : KB^2$$
$$= AS^2 : SQ^2;$$
therefore $\quad SQ^2 : SO^2 = AS^2 : AS \cdot SC$
$$= SQ^2 : SQ \cdot QM,$$
and accordingly $\quad SO^2 = SQ \cdot QM.$

Add SQ^2 to each side, and we have
$$SO^2 + SQ^2 = SQ \cdot SM.$$

Therefore, from above, we have
$HA : AS = MS^2 : (SO^2 + SQ^2)$
$\quad = MN^2 : (OP^2 + QR^2)$
$\quad =$ (circle, diam. MN):(circle, diam. OP + circle, diam. QR)

That is,

$HA : AS =$ (circle in cylinder) : (circle in spheroid + circle in cone).

Therefore the circle in the cylinder, in the place where it is, is in equilibrium, about A, with the circle in the spheroid and the circle in the cone together, if both the latter circles are placed with their centres of gravity at H.

Similarly for the three corresponding sections made by a plane perpendicular to AC and passing through any other straight line in the parallelogram LF parallel to EF.

If we deal in the same way with all the sets of three circles in which planes perpendicular to AC cut the cylinder, the spheroid and the cone, and which make up those figures respectively, it follows that the cylinder, in the place where it is, will be in equilibrium about A with the spheroid and the cone together, when both are placed with their centres of gravity at H.

Therefore, since K is the centre of gravity of the cylinder,

$$HA : AK = \text{(cylinder)} : (\text{spheroid} + \text{cone } AEF).$$

But $HA = 2AK$;

therefore $\quad$ cylinder $= 2$ (spheroid + cone AEF).

And $\quad$ cylinder $= 3$ (cone AEF); $\quad$ [Eucl. XII. 10]

therefore $\quad$ cone $AEF = 2$ (spheroid).

But, since $EF = 2BD$,

cone $AEF = 8$ (cone ABD);

therefore $\quad$ spheroid $= 4$ (cone ABD),

and $\quad$ half the spheroid $= 2$ (cone ABD).

Through B, D draw VBW, XDY parallel to AC; and imagine a cylinder which has AC for axis and the circles on VX, WY as diameters for bases.

Then $\quad$ cylinder $VY = 2$ (cylinder VD)

$= 6$ (cone ABD)

$= \frac{3}{2}$ (spheroid), from above.

Q.E.D.

Proposition 4.

Any segment of a right-angled conoid (i.e. a paraboloid of revolution) cut off by a plane at right angles to the axis is $1\frac{1}{2}$ times the cone which has the same base and the same axis as the segment.

This can be investigated by our method, as follows.

Let a paraboloid of revolution be cut by a plane through the axis in the parabola BAC;

and let it also be cut by another plane at right angles to the axis and intersecting the former plane in BC. Produce DA, the axis of the segment, to H, making HA equal to AD.

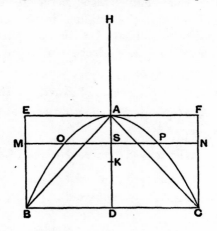

Imagine that HD is the bar of a balance, A being its middle point.

The base of the segment being the circle on BC as diameter and in a plane perpendicular to AD,
imagine (1) a cone drawn with the latter circle as base and A as vertex, and (2) a cylinder with the same circle as base and AD as axis.

In the parallelogram EC let any straight line MN be drawn parallel to BC, and through MN let a plane be drawn at right angles to AD; this plane will cut the cylinder in a circle with diameter MN and the paraboloid in a circle with diameter OP.

Now, BAC being a parabola and BD, OS ordinates,

$$DA : AS = BD^2 : OS^2,$$

or

$$HA : AS = MS^2 : SO^2.$$

Therefore

$$HA : AS = (\text{circle, rad. } MS) : (\text{circle, rad. } OS)$$
$$= (\text{circle in cylinder}) : (\text{circle in paraboloid}).$$

Therefore the circle in the cylinder, in the place where it is, will be in equilibrium, about A, with the circle in the paraboloid, if the latter is placed with its centre of gravity at H.

Similarly for the two corresponding circular sections made by a plane perpendicular to AD and passing through any other straight line in the parallelogram which is parallel to BC.

Therefore, as usual, if we take all the circles making up the whole cylinder and the whole segment and treat them in the same way, we find that the cylinder, in the place where it is, is in equilibrium about A with the segment placed with its centre of gravity at H.

If K is the middle point of AD, K is the centre of gravity of the cylinder;

therefore $HA : AK = (\text{cylinder}) : (\text{segment}).$

Therefore cylinder $= 2$ (segment).

And cylinder $= 3$ (cone ABC); [Eucl. XII. 10]

therefore segment $= \frac{3}{2}$ (cone ABC).

Proposition 5.

The centre of gravity of a segment of a right-angled conoid (i.e. a paraboloid of revolution) cut off by a plane at right angles to the axis is on the straight line which is the axis of the segment, and divides the said straight line in such a way that the portion of it adjacent to the vertex is double of the remaining portion.

This can be investigated by the method, as follows.

Let a paraboloid of revolution be cut by a plane through the axis in the parabola BAC;

and let it also be cut by another plane at right angles to the axis and intersecting the former plane in BC.

Produce DA, the axis of the segment, to H, making HA equal to AD; and imagine DH to be the bar of a balance, its middle point being A.

The base of the segment being the circle on BC as diameter and in a plane perpendicular to AD,
imagine a cone with this circle as base and A as vertex, so that AB, AC are generators of the cone.

In the parabola let any double ordinate OP be drawn meeting AB, AD, AC in Q, S, R respectively.

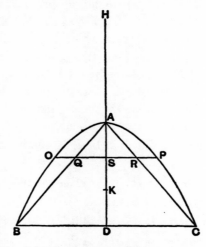

Now, from the property of the parabola,
$$BD^2 : OS^2 = DA : AS$$
$$= BD : QS$$
$$= BD^2 : BD \cdot QS.$$

Therefore $OS^2 = BD \cdot QS$,
or $BD : OS = OS : QS$,
whence $BD : QS = OS^2 : QS^2$.
But $BD : QS = AD : AS$
$$= HA : AS.$$
Therefore $HA : AS = OS^2 : QS^2$
$$= OP^2 : QR^2.$$

If now through OP a plane be drawn at right angles to AD, this plane cuts the paraboloid in a circle with diameter OP and the cone in a circle with diameter QR.

We see therefore that

$$HA : AS = (\text{circle, diam. } OP) : (\text{circle, diam. } QR)$$
$$= (\text{circle in paraboloid}) : (\text{circle in cone});$$

and the circle in the paraboloïd, in the place where it is, is in equilibrium about A with the circle in the cone placed with its centre of gravity at H.

Similarly for the two corresponding circular sections made by a plane perpendicular to AD and passing through any other ordinate of the parabola.

Dealing therefore in the same way with all the circular sections which make up the whole of the segment of the paraboloid and the cone respectively, we see that the segment of the paraboloid, in the place where it is, is in equilibrium about A with the cone placed with its centre of gravity at H.

Now, since A is the centre of gravity of the whole system as placed, and the centre of gravity of part of it, namely the cone, as placed, is at H, the centre of gravity of the rest, namely the segment, is at a point K on HA produced such that

$$HA : AK = (\text{segment}) : (\text{cone}).$$

But $\quad$ segment $= \frac{3}{2}$ (cone). $\quad$ [Prop. 4]

Therefore $\quad HA = \frac{3}{2} AK$;

that is, K divides AD in such a way that $AK = 2KD$.

Proposition 6.

The centre of gravity of any hemisphere [is on the straight line which] is its axis, and divides the said straight line in such a way that the portion of it adjacent to the surface of the hemisphere has to the remaining portion the ratio which 5 has to 3.

Let a sphere be cut by a plane through its centre in the circle $ABCD$;

let AC, BD be perpendicular diameters of this circle,

and through BD let a plane be drawn at right angles to AC.

The latter plane will cut the sphere in a circle on BD as diameter.

Imagine a cone with the latter circle as base and A as vertex.

Produce CA to H, making AH equal to CA, and let HC be regarded as the bar of a balance, A being its middle point.

N
H
M
A
O
P
Q
E
W
R
V
B
G
D
C

In the semicircle BAD, let any straight line OP be drawn parallel to BD and cutting AC in E and the two generators AB, AD of the cone in Q, R respectively. Join AO.

Through OP let a plane be drawn at right angles to AC;

this plane will cut the hemisphere in a circle with diameter OP and the cone in a circle with diameter QR.

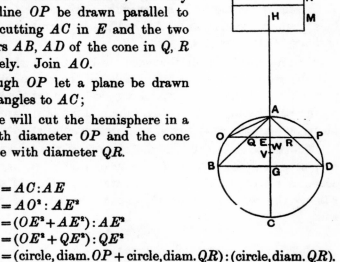

Now

$$\begin{aligned} HA:AE &= AC:AE \\ &= AO^2 : AE^2 \\ &= (OE^2+AE^2):AE^2 \\ &= (OE^2+QE^2):QE^2 \\ &= (\text{circle, diam. } OP + \text{circle, diam. } QR) : (\text{circle, diam. } QR). \end{aligned}$$

Therefore the circles with diameters OP, QR, in the places where they are, are in equilibrium about A with the circle with diameter QR if the latter is placed with its centre of gravity at H.

And, since the centre of gravity of the two circles with diameters OP, QR taken together, in the place where they are, is......

[There is a lacuna here; but the proof can easily be completed on the lines of the corresponding but more difficult case in Prop. 8.

We proceed thus from the point where the circles with diameters OP, QR, in the place where they are, balance, about A, the circle with diameter QR placed with its centre of gravity at H.

A similar relation holds for all the other sets of circular sections made by other planes passing through points on AG and at right angles to AG.

Taking then all the circles which fill up the hemisphere BAD and the cone ABD respectively, we find that

the hemisphere BAD and the cone ABD, in the places where they are, together balance, about A, a cone equal to ABD placed with its centre of gravity at H.

Let the cylinder $M+N$ be equal to the cone ABD.

Then, since the cylinder $M+N$ placed with its centre of gravity at H balances the hemisphere BAD and the cone ABD in the places where they are,

suppose that the portion M of the cylinder, placed with its centre of gravity at H, balances the cone ABD (alone) in the place where it is; therefore the portion N of the cylinder placed with its centre of gravity at H balances the hemisphere (alone) in the place where it is.

Now the centre of gravity of the cone is at a point V such that $AG = 4GV$;

therefore, since M at H is in equilibrium with the cone,

$$M : (\text{cone}) = \tfrac{3}{4}AG : HA = \tfrac{3}{8}AC : AC,$$

whence $\qquad M = \tfrac{3}{8}(\text{cone})$.

But $M + N = (\text{cone})$; therefore $N = \tfrac{5}{8}(\text{cone})$.

Now let the centre of gravity of the hemisphere be at W, which is somewhere on AG.

Then, since N at H balances the hemisphere alone,

(hemisphere) $: N = HA : AW$.

But the hemisphere $BAD =$ twice the cone ABD;

[*On the Sphere and Cylinder* I. 34 and Prop. 2 above]

and $N = \tfrac{5}{8}(\text{cone})$, from above.

Therefore $\qquad 2 : \tfrac{5}{8} = HA : AW$

$\qquad\qquad\qquad\quad = 2AG : AW$,

whence $AW = \tfrac{5}{8}AG$, so that W divides AG in such a way that

$$AW : WG = 5 : 3.]$$

Proposition 7.

We can also investigate by the same method the theorem that

[*Any segment of a sphere has*] *to the cone* [*with the same base and height the ratio which the sum of the radius of the sphere and the height of the complementary segment has to the height of the complementary segment.*]

[There is a lacuna here; but all that is missing is the construction, and the construction is easily understood by

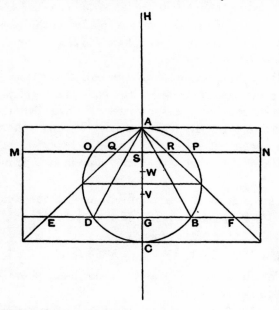

means of the figure. *BAD* is of course the segment of the sphere the volume of which is to be compared with the volume of a cone with the same base and height.]

The plane drawn through *MN* and at right angles to *AC* will cut the cylinder in a circle with diameter *MN*, the segment of the sphere in a circle with diameter *OP*, and the cone on the base *EF* in a circle with diameter *QR*.

In the same way as before [cf. Prop. 2] we can prove that the circle with diameter *MN*, in the place where it is, is in equilibrium about A with the two circles with diameters OP, QR if these circles are both moved and placed with their centres of gravity at H.

The same thing can be proved of all sets of three circles in which the cylinder, the segment of the sphere, and the cone with the common height AG are all cut by any plane perpendicular to AC.

Since then the sets of circles make up the whole cylinder, the whole segment of the sphere and the whole cone respectively, it follows that the cylinder, in the place where it is, is in equilibrium about A with the sum of the segment of the sphere and the cone if both are placed with their centres of gravity at H.

Divide AG at W, V in such a way that

$$AW = WG, \quad AV = 3VG.$$

Therefore W will be the centre of gravity of the cylinder, and V will be the centre of gravity of the cone.

Since, now, the bodies are in equilibrium as described,

$$(\text{cylinder}) : (\text{cone } AEF + \text{segment } BAD \text{ of sphere}) = HA : AW.$$

....

[The rest of the proof is lost; but it can easily be supplied thus.

We have

$$(\text{cone } AEF + \text{segmt. } BAD) : (\text{cylinder}) = AW : AC$$
$$= AW \,.\, AC : AC^2.$$

But $$(\text{cylinder}) : (\text{cone } AEF) = AC^2 : \tfrac{1}{3}EG^2$$
$$= AC^2 : \tfrac{1}{3}AG^2.$$

Therefore, *ex aequali*,

$$(\text{cone } AEF + \text{segmt. } BAD) : (\text{cone } AEF) = AW \,.\, AC : \tfrac{1}{3}AG^2$$
$$= \tfrac{1}{2}AC : \tfrac{1}{3}AG,$$

whence $$(\text{segmt. } BAD) : (\text{cone } AEF) = (\tfrac{1}{2}AC - \tfrac{1}{3}AG) : \tfrac{1}{3}AG.$$

Again $$(\text{cone } AEF) : (\text{cone } ABD) = EG^2 : DG^2$$
$$= AG^2 : AG \,.\, GC$$
$$= AG : GC$$
$$= \tfrac{1}{3}AG : \tfrac{1}{3}GC.$$

Therefore, *ex aequali*,

$$(\text{segment } BAD) : (\text{cone } ABD) = (\tfrac{1}{2}AC - \tfrac{1}{3}AG) : \tfrac{1}{3}GC$$
$$= (\tfrac{3}{2}AC - AG) : GC$$
$$= (\tfrac{1}{2}AC + GC) : GC.$$

Q.E.D.]

Proposition 8.

[The enunciation, the setting-out, and a few words of the construction are missing.

The enunciation however can be supplied from that of Prop. 9, with which it must be identical except that it cannot refer to "*any* segment," and the presumption therefore is that the proposition was enunciated with reference to one kind of segment only, i.e. either a segment greater than a hemisphere or a segment less than a hemisphere.

Heiberg's figure corresponds to the case of a segment greater than a hemisphere. The segment investigated is of course the segment *BAD*. The setting-out and construction are self-evident from the figure.]

N
M
H
A
K
P
L
R
Q
W
V
E
B
G
D
F
C
O

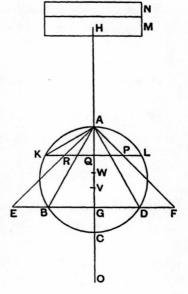

Produce *AC* to *H*, *O*, making *HA* equal to *AC* and *CO* equal to the radius of the sphere;

and let *HC* be regarded as the bar of a balance, the middle point being *A*.

In the plane cutting off the segment describe a circle with *G* as centre and radius (*GE*) equal to *AG*; and on this circle as base, and with *A* as vertex, let a cone be described. *AE*, *AF* are generators of this cone.

Draw *KL*, through any point *Q* on *AG*, parallel to *EF* and cutting the segment in *K*, *L*, and *AE*, *AF* in *R*, *P* respectively. Join *AK*.

Now $HA : AQ = CA : AQ$
$= AK^2 : AQ^2$
$= (KQ^2 + QA^2) : QA^2$
$= (KQ^2 + PQ^2) : PQ^2$
= (circle, diam. KL + circle, diam. PR)
: (circle, diam. PR).

Imagine a circle equal to the circle with diameter PR placed with its centre of gravity at H;

therefore the circles on diameters KL, PR, in the places where they are, are in equilibrium about A with the circle with diameter PR placed with its centre of gravity at H.

Similarly for the corresponding circular sections made by any other plane perpendicular to AG.

Therefore, taking all the circular sections which make up the segment ABD of the sphere and the cone AEF respectively, we find that the segment ABD of the sphere and the cone AEF, in the places where they are, are in equilibrium with the cone AEF assumed to be placed with its centre of gravity at H.

Let the cylinder $M + N$ be equal to the cone AEF which has A for vertex and the circle on EF as diameter for base.

Divide AG at V so that $AG = 4VG$;
therefore V is the centre of gravity of the cone AEF; "for this has been proved before*."

Let the cylinder $M + N$ be cut by a plane perpendicular to the axis in such a way that the cylinder M (alone), placed with its centre of gravity at H, is in equilibrium with the cone AEF.

Since $M + N$ suspended at H is in equilibrium with the segment ABD of the sphere and the cone AEF in the places where they are,

while M, also at H, is in equilibrium with the cone AEF in the place where it is, it follows that

N at H is in equilibrium with the segment ABD of the sphere in the place where it is.

* Cf. note on p. 15 above.

Now (segment ABD of sphere) : (cone ABD)
$$= OG : GC;$$
"for this is already proved" [Cf. *On the Sphere and Cylinder* II. 2 Cor. as well as Prop. 7 *ante*].

And (cone ABD) : (cone AEF)
$$\begin{aligned} &= (\text{circle, diam. } BD) : (\text{circle, diam. } EF) \\ &= BD^2 : EF^2 \\ &= BG^2 : GE^2 \\ &= CG \,.\, GA : GA^2 \\ &= CG : GA. \end{aligned}$$

Therefore, *ex aequali*,
$$(\text{segment } ABD \text{ of sphere}) : (\text{cone } AEF) = OG : GA.$$

Take a point W on AG such that
$$AW : WG = (GA + 4GC) : (GA + 2GC).$$
We have then, inversely,
$$GW : WA = (2GC + GA) : (4GC + GA),$$
and, *componendo*,
$$GA : AW = (6GC + 2GA) : (4GC + GA).$$

But $GO = \frac{1}{4}(6GC + 2GA)$, [for $GO - GC = \frac{1}{2}(CG + GA)$]
and $CV = \frac{1}{4}(4GC + GA)$;
therefore $GA : AW = OG : CV$,
and, alternately and inversely,
$$OG : GA = CV : WA.$$

It follows, from above, that
$$(\text{segment } ABD \text{ of sphere}) : (\text{cone } AEF) = CV : WA.$$

Now, since the cylinder M with its centre of gravity at H is in equilibrium about A with the cone AEF with its centre of gravity at V,
$$\begin{aligned} (\text{cone } AEF) : (\text{cylinder } M) &= HA : AV \\ &= CA : AV; \end{aligned}$$
and, since the cone AEF = the cylinder $M + N$, we have, *dividendo* and *invertendo*,
$$(\text{cylinder } M) : (\text{cylinder } N) = AV : CV.$$

Hence, *componendo*,

$$(\text{cone } AEF) : (\text{cylinder } N) = CA : CV^{*}$$
$$= HA : CV.$$

But it was proved that

$$(\text{segment } ABD \text{ of sphere}) : (\text{cone } AEF) = CV : WA;$$

therefore, *ex aequali*,

$$(\text{segment } ABD \text{ of sphere}) : (\text{cylinder } N) = HA : AW.$$

And it was above proved that the cylinder N at H is in equilibrium about A with the segment ABD, in the place where it is;

therefore, since H is the centre of gravity of the cylinder N, W is the centre of gravity of the segment ABD of the sphere.

Proposition 9.

In the same way we can investigate the theorem that

The centre of gravity of any segment of a sphere is on the straight line which is the axis of the segment, and divides this straight line in such a way that the part of it adjacent to the vertex of the segment has to the remaining part the ratio which the sum of the axis of the segment and four times the axis of the complementary segment has to the sum of the axis of the segment and double the axis of the complementary segment.

[As this theorem relates to "*any* segment" but states the same result as that proved in the preceding proposition, it follows that Prop. 8 must have related to one kind of segment, either a segment greater than a semicircle (as in Heiberg's figure of Prop. 8) or a segment less than a semicircle; and the present proposition completed the proof for both kinds of segments. It would only require a slight change in the figure, in any case.]

Proposition 10.

By this method too we can investigate the theorem that

[*A segment of an obtuse-angled conoid* (*i.e. a hyperboloid of revolution*) *has to the cone which has*] *the same base* [*as the*

* Archimedes arrives at this result in a very roundabout way, seeing that it could have been obtained at once *convertendo*. Cf. Euclid x. 14.

segment and equal height the same ratio as the sum of the axis of the segment and three times] *the " annex to the axis" (i.e. half the transverse axis of the hyperbolic section through the axis of the hyperboloid or, in other words, the distance between the vertex of the segment and the vertex of the enveloping cone) has to the sum of the axis of the segment and double of the "annex"* *
[this is the theorem proved in On Conoids and Spheroids, Prop. 25], "and also many other theorems, which, as the method has been made clear by means of the foregoing examples, I will omit, in order that I may now proceed to compass the proofs of the theorems mentioned above."

Proposition 11.

If in a right prism with square bases a cylinder be inscribed having its bases in opposite square faces and touching with its surface the remaining four parallelogrammic faces, and if through the centre of the circle which is the base of the cylinder and one side of the opposite square face a plane be drawn, the figure cut off by the plane so drawn is one sixth part of the whole prism.

"This can be investigated by the method, and, when it is set out, I will go back to the proof of it by geometrical considerations."

[The investigation by the mechanical method is contained in the two Propositions, 11, 12. Prop. 13 gives another solution which, although it contains no mechanics, is still of the character which Archimedes regards as inconclusive, since it assumes that the solid is actually *made up* of parallel plane sections and that an auxiliary parabola is actually *made up* of parallel straight lines in it. Prop. 14 added the conclusive geometrical proof.]

Let there be a right prism with a cylinder inscribed as stated.

---

* The text has "triple" ($\tau\rho\iota\pi\lambda\alpha\sigma\iota\alpha\nu$) in the last line instead of "double." As there is a considerable lacuna before the last few lines, a theorem about the centre of gravity of a segment of a hyperboloid of revolution may have fallen out.

Let the prism be cut through the axis of the prism and cylinder by a plane perpendicular to the plane which cuts off the portion of the cylinder; let this plane make, as section, the parallelogram AB, and let it cut the plane cutting off the portion of the cylinder (which plane is perpendicular to AB) in the straight line BC.

D
B
U
G
F
E
I
A
C
V
Y

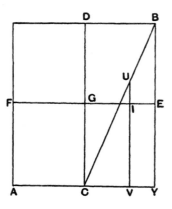

Let CD be the axis of the prism and cylinder, let EF bisect it at right angles, and through EF let a plane be drawn at right angles to CD; this plane will cut the prism in a square and the cylinder in a circle.

Let MN be the square and $OPQR$ the circle, and let the circle touch the sides of the square in O, P, Q, R [F, E in the first figure are identical with O, Q respectively]. Let H be the centre of the circle.

R
N
L
T
O
H
W
Q
K
S
M
P

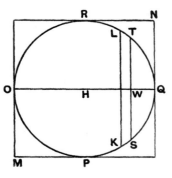

Let KL be the intersection of the plane through EF perpendicular to the axis of the cylinder and the plane cutting off the portion of the cylinder; KL is bisected by OHQ [and passes through the middle point of HQ].

Let any chord of the circle, as ST, be drawn perpendicular to HQ, meeting HQ in W;

and through ST let a plane be drawn at right angles to OQ and produced on both sides of the plane of the circle $OPQR$.

The plane so drawn will cut the half cylinder having the semicircle PQR for section and the axis of the prism for height in a parallelogram, one side of which is equal to ST and another is a generator of the cylinder; and it will also cut the portion of the cylinder cut off in a parallelogram, one side of which is equal to ST and the other is equal and parallel to UV (in the first figure).

UV will be parallel to BY and will cut off, along EG in the parallelogram DE, the segment EI equal to QW.

Now, since EC is a parallelogram, and VI is parallel to GC,

$EG : GI = YC : CV$
$\qquad = BY : UV$
$\qquad = (\square$ in half cyl.$) : (\square$ in portion of cyl.$)$.

And $EG = HQ$, $GI = HW$, $QH = OH$;

therefore $\quad OH : HW = (\square$ in half cyl.$) : (\square$ in portion$)$.

Imagine that the parallelogram in the portion of the cylinder is moved and placed at O so that O is its centre of gravity, and that OQ is the bar of a balance, H being its middle point.

Then, since W is the centre of gravity of the parallelogram in the half cylinder, it follows from the above that the parallelogram in the half cylinder, in the place where it is, with its centre of gravity at W, is in equilibrium about H with the parallelogram in the portion of the cylinder when placed with its centre of gravity at O.

Similarly for the other parallelogrammic sections made by any plane perpendicular to OQ and passing through any other chord in the semicircle PQR perpendicular to OQ.

If then we take all the parallelograms making up the half cylinder and the portion of the cylinder respectively, it follows that the half cylinder, in the place where it is, is in equilibrium about H with the portion of the cylinder cut off when the latter is placed with its centre of gravity at O.

Proposition 12.

Let the parallelogram (square) MN perpendicular to the axis, with the circle $OPQR$ and its diameters OQ, PR, be drawn separately.

THE METHOD

Join HG, HM, and through them draw planes at right angles to the plane of the circle, producing them on both sides of that plane.

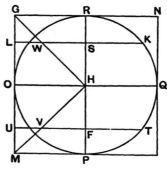

This produces a prism with triangular section GHM and height equal to the axis of the cylinder; this prism is $\frac{1}{4}$ of the original prism circumscribing the cylinder.

Let LK, UT be drawn parallel to OQ and equidistant from it, cutting the circle in K, T, RP in S, F, and GH, HM in W, V respectively.

Through LK, UT draw planes at right angles to PR, producing them on both sides of the plane of the circle; these planes produce as sections in the half cylinder PQR and in the prism GHM four parallelograms in which the heights are equal to the axis of the cylinder, and the other sides are equal to KS, TF, LW, UV respectively..............

....

[The rest of the proof is missing, but, as Zeuthen says*, the result obtained and the method of arriving at it are plainly indicated by the above.

Archimedes wishes to prove that the half cylinder PQR, in the place where it is, balances the prism GHM, in the place where it is, about H as fixed point.

He has first to prove that the elements (1) the parallelogram with side $= KS$ and (2) the parallelogram with side $= LW$, in the places where they are, balance about S, or, in other words that the straight lines SK, LW, in the places where they are, balance about S.

Now (radius of circle $OPQR)^2 = SK^2 + SH^2$,
or $SL^2 = SK^2 + SW^2$.
Therefore $LS^2 - SW^2 = SK^2$,
and accordingly $(LS + SW) \cdot LW = SK^2$,
whence $\frac{1}{2}(LS + SW) : \frac{1}{2}SK = SK : LW$.

* Zeuthen in *Bibliotheca Mathematica* vII$_3$, 1906–7, pp. 352–3.

And $\frac{1}{2}(LS+SW)$ is the distance of the centre of gravity of LW from S,

while $\frac{1}{2}SK$ is the distance of the centre of gravity of SK from S.

Therefore SK and LW, in the places where they are, balance about S.

Similarly for the corresponding parallelograms.

Taking *all* the parallelogrammic elements in the half cylinder and prism respectively, we find that

the half cylinder PQR and the prism GHM, in the places where they are respectively, balance about H.

From this result and that of Prop. 11 we can at once deduce the volume of the portion cut off from the cylinder. For in Prop. 11 the portion of the cylinder, placed with its centre of gravity at O, is shown to balance (about H) the half-cylinder in the place where it is. By Prop. 12 we may substitute for the half-cylinder in the place where it is the prism GHM of that proposition turned the opposite way relatively to RP. The centre of gravity of the prism as thus placed is at a point (say Z) on HQ such that $HZ = \frac{2}{3}HQ$.

Therefore, assuming the prism to be applied at its centre of gravity, we have

(portion of cylinder) : (prism) $= \frac{2}{3}HQ : OH$
$= 2 : 3$;

therefore (portion of cylinder) $= \frac{2}{3}$ (prism GHM)
$= \frac{1}{3}$ (original prism).

Note. This proposition of course solves the problem of finding the centre of gravity of a half cylinder or, in other words, of a semicircle.

For the triangle GHM in the place where it is balances, about H, the semicircle PQR in the place where it is.

If then X is the point on HQ which is the centre of gravity of the semicircle,

$$\tfrac{2}{3}HO . (\triangle GHM) = HX . (\text{semicircle } PQR),$$
or $\qquad \tfrac{2}{3}HO . HO^2 = HX . \tfrac{1}{2}\pi . HO^2;$

that is, $\qquad HX = \dfrac{4}{3\pi} . HQ.$]

Proposition 13.

Let there be a right prism with square bases, one of which is *ABCD*;

in the prism let a cylinder be inscribed, the base of which is the circle *EFGH* touching the sides of the square *ABCD* in *E, F, G, H*.

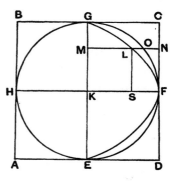

Through the centre and through the side corresponding to *CD* in the square face *opposite* to *ABCD* let a plane be drawn; this will cut off a prism equal to ¼ of the original prism and formed by three parallelograms and two triangles, the triangles forming opposite faces.

In the semicircle *EFG* describe the parabola which has *FK* for axis and passes through *E, G*;
draw *MN* parallel to *KF* meeting *GE* in *M*, the parabola in *L*, the semicircle in *O* and *CD* in *N*.
Then $MN \cdot NL = NF^2$;
"for this is clear." [Cf. Apollonius, *Conics* I. 11]
[The parameter is of course equal to *GK* or *KF*.]
Therefore $MN : NL = GK^2 : LS^2$.

Through *MN* draw a plane at right angles to *EG*; this will produce as sections (1) in the prism cut off from the whole prism a right-angled triangle, the base of which is *MN*, while the perpendicular is perpendicular at *N* to the plane *ABCD* and equal to the axis of the cylinder, and the hypotenuse is in the plane cutting the cylinder, and (2) in the portion of the cylinder cut off a right-angled triangle the base of which is *MO*, while the perpendicular is the generator of the cylinder perpendicular at *O* to the plane *KN*, and the hypotenuse is.....................

....

[There is a lacuna here, to be supplied as follows.
Since $MN : NL = GK^2 : LS^2$
$= MN^2 : LS^2$,
it follows that $MN : ML = MN^2 : (MN^2 - LS^2)$
$= MN^2 : (MN^2 - MK^2)$
$= MN^2 : MO^2$.

But the triangle (1) in the prism is to the triangle (2) in the portion of the cylinder in the ratio of $MN^2 : MO^2$.
Therefore
($\triangle$ in prism) : ($\triangle$ in portion of cylinder)
$= MN : ML$
$=$ (straight line in rect. DG) : (straight line in parabola).

We now take all the corresponding elements in the prism, the portion of the cylinder, the rectangle DG and the parabola EFG respectively ;]
and it will follow that
(all the $\triangle$s in prism) : (all the $\triangle$s in portion of cylinder)
$=$ (all the str. lines in $\square\ DG$)
: (all the straight lines between parabola and EG).

But the prism is made up of the triangles in the prism, [the portion of the cylinder is made up of the triangles in it], the parallelogram DG of the straight lines in it parallel to KF, and the parabolic segment of the straight lines parallel to KF intercepted between its circumference and EG;
therefore   (prism) : (portion of cylinder)
$= (\square\ GD)$: (parabolic segment EFG).

But $\square\ GD = \frac{3}{2}$ (parabolic segment EFG);
"for this is proved in my earlier treatise."

[*Quadrature of Parabola*]

Therefore   prism $= \frac{3}{2}$ (portion of cylinder).

If then we denote the portion of the cylinder by 2, the prism is 3, and the original prism circumscribing the cylinder is 12 (being 4 times the other prism);
therefore the portion of the cylinder $= \frac{1}{6}$ (original prism).

Q.E.D.

THE METHOD

[The above proposition and the next are peculiarly interesting for the fact that the parabola is an auxiliary curve introduced for the sole purpose of analytically reducing the required cubature to the known quadrature of the parabola.]

Proposition 14.

Let there be a right prism with square bases [and a cylinder inscribed therein having its base in the square $ABCD$ and touching its sides at E, F, G, H;
let the cylinder be cut by a plane through EG and the side corresponding to CD in the square face opposite to $ABCD$].

This plane cuts off from the prism a prism, and from the cylinder a portion of it.

It can be proved that the portion of the cylinder cut off by the plane is $\frac{1}{6}$ of the whole prism.

But we will first prove that it is possible to inscribe in the portion cut off from the cylinder, and to circumscribe about it, solid figures made up of prisms which have equal height and similar triangular bases, in such a way that the circumscribed figure exceeds the inscribed by less than any assigned magnitude..
....

But it was proved that
 (prism cut off by oblique plane)
  $< \frac{3}{2}$ (figure inscribed in portion of cylinder).
Now
 (prism cut off) : (inscribed figure)
  $= \Box\, DG : (\Box\text{s inscribed in parabolic segment})$;
therefore $\Box\, DG < \frac{3}{2} (\Box\text{s in parabolic segment})$:
which is impossible, since "it has been proved elsewhere" that the parallelogram DG is $\frac{3}{2}$ of the parabolic segment.

Consequently........................
....not greater.
....

And (all the prisms in prism cut off)
: (all prisms in circumscr. figure)
= (all ▱s in ▱ DG)
: (all ▱s in fig. circumscr. about parabolic segmt.);
therefore
(prism cut off) : (figure circumscr. about portion of cylinder)
= (▱ DG) : (figure circumscr. about parabolic segment).

But the prism cut off by the oblique plane is > $\frac{3}{2}$ of the solid figure circumscribed about the portion of the cylinder.....................

....

[There are large gaps in the exposition of this geometrical proof, but the way in which the method of exhaustion was applied, and the parallelism between this and other applications of it, are clear. The first fragment shows that solid figures made up of prisms were circumscribed and inscribed to the portion of the cylinder. The parallel triangular faces of these prisms were perpendicular to GE in the figure of Prop. 13; they divided GE into equal portions of the requisite smallness; each section of the portion of the cylinder by such a plane was a triangular face common to an inscribed and a circumscribed right prism. The planes also produced prisms in the prism cut off by the same oblique plane as cuts off the portion of the cylinder and standing on GD as base.

The number of parts into which the parallel planes divided GE was made great enough to secure that the circumscribed figure exceeded the inscribed figure by less than a small assigned magnitude.

The second part of the proof began with the assumption that the portion of the cylinder is > $\frac{2}{3}$ of the prism cut off; and this was proved to be impossible, by means of the use of the auxiliary parabola and the proportion

$$MN : ML = MN^2 : MO^2$$

which are employed in Prop. 13.

We may supply the missing proof as follows*.

In the accompanying figure are represented (1) the first

C′
N′
D′
R
C
N
F
D
O
G
M
K
E
B
H
A

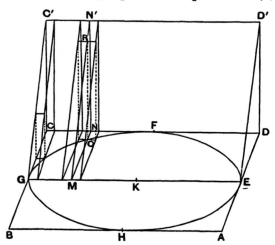

element-prism circumscribed to the portion of the cylinder, (2) two element-prisms adjacent to the ordinate *OM*, of which that on the left is circumscribed and that on the right (equal to the other) inscribed, (3) the corresponding element-prisms forming part of the prism cut off (*CC′GEDD′*) which is ¼ of the original prism.

G
C
M
L
O
N
K
F
E
D

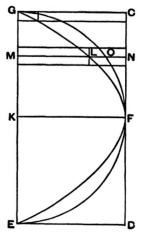

In the second figure are shown element-rectangles circumscribed and inscribed to the auxiliary parabola, which rectangles correspond exactly to the circumscribed and inscribed element-prisms represented in the first figure (the length of *GM* is the same in both figures, and the breadths of the element-rectangles are the same as the heights of the element-prisms);

* It is right to mention that this has already been done by Th. Reinach in his version of the treatise ("Un Traité de Géométrie inédit d'Archimède" in *Revue générale des sciences pures et appliquées*, 30 Nov. and 15 Dec. 1907); but I prefer my own statement of the proof.

the corresponding element-rectangles forming part of the rectangle GD are similarly shown.

For convenience we suppose that GE is divided into an even number of equal parts, so that GK contains an integral number of these parts.

For the sake of brevity we will call each of the two element-prisms of which OM is an edge "el. prism (O)" and each of the element-prisms of which MNN' is a common face "el. prism (N)." Similarly we will use the corresponding abbreviations "el. rect. (L)" and "el. rect. (N)" for the corresponding elements in relation to the auxiliary parabola as shown in the second figure.

Now it is easy to see that the figure made up of all the inscribed prisms is less than the figure made up of the circumscribed prisms by twice the final circumscribed prism adjacent to FK, i.e. by twice "el. prism (N)"; and, as the height of this prism may be made as small as we please by dividing GK into sufficiently small parts, it follows that inscribed and circumscribed solid figures made up of element-prisms can be drawn differing by less than any assigned solid figure.

(1) Suppose, if possible, that

$$\text{(portion of cylinder)} > \tfrac{2}{3}\,\text{(prism cut off)},$$

or

$$\text{(prism cut off)} < \tfrac{3}{2}\,\text{(portion of cylinder)}.$$

Let $\text{(prism cut off)} = \tfrac{3}{2}\,(\text{portion of cylinder} - X)$, say.

Construct circumscribed and inscribed figures made up of element-prisms, such that

$$\text{(circumscr. fig.)} - \text{(inscr. fig.)} < X.$$

Therefore $\qquad \text{(inscr. fig.)} > (\text{circumscr. fig.} - X),$

and *a fortiori* $\qquad > (\text{portion of cyl.} - X).$

It follows that

$$\text{(prism cut off)} < \tfrac{3}{2}\,\text{(inscribed figure)}.$$

Considering now the element-prisms in the prism cut off and those in the inscribed figure respectively, we have

$$\begin{aligned} \text{el. prism } (N) : \text{el. prism } (O) &= MN^2 : MO^2 \\ &= MN : ML \quad \text{[as in Prop. 13]} \\ &= \text{el. rect. } (N) : \text{el. rect. } (L). \end{aligned}$$

It follows that

$$\Sigma \{\text{el. prism } (N)\} : \Sigma \{\text{el. prism } (O)\} = \Sigma \{\text{el. rect. } (N)\} : \Sigma \{\text{el. rect. } (L)\}.$$

(There are really two more prisms and rectangles in the first and third than there are in the second and fourth terms respectively; but this makes no difference because the first and third terms may be multiplied by a common factor as $n/(n-2)$ without affecting the truth of the proportion. Cf. the proposition from *On Conoids and Spheroids* quoted on p. 15 above.)

Therefore

$$(\text{prism cut off}) : (\text{figure inscr. in portion of cyl.}) = (\text{rect. } GD) : (\text{fig. inscr. in parabola}).$$

But it was proved above that

$$(\text{prism cut off}) < \tfrac{3}{2}\,(\text{fig. inscr. in portion of cyl.});$$

therefore $(\text{rect. } GD) < \tfrac{3}{2}\,(\text{fig. inscr. in parabola})$,

and, *a fortiori*,

$$(\text{rect. } GD) < \tfrac{3}{2}\,(\text{parabolic segmt.}):$$

which is impossible, since

$$(\text{rect. } GD) = \tfrac{3}{2}\,(\text{parabolic segmt.}).$$

Therefore

(portion of cyl.) is *not* greater than $\tfrac{2}{3}$ (prism cut off).

(2) In the second lacuna must have come the beginning of the next *reductio ad absurdum* demolishing the other possible assumption that the portion of the cylinder is $< \tfrac{2}{3}$ of the prism cut off.

In this case our assumption is that

$$(\text{prism cut off}) > \tfrac{3}{2}\,(\text{portion of cylinder});$$

and we circumscribe and inscribe figures made up of element-prisms, such that

$$(\text{prism cut off}) > \tfrac{3}{2}\,(\text{fig. circumscr. about portion of cyl.}).$$

We now consider the element-prisms in the prism cut off and in the circumscribed figure respectively, and the same argument as above gives

(prism cut off) : (fig. circumscr. about portion of cyl.)
= (rect. GD) : (fig. circumscr. about parabola),

whence it follows that

(rect. GD) > $\frac{2}{3}$ (fig. circumscribed about parabola),

and, *a fortiori*,

(rect. GD) > $\frac{2}{3}$ (parabolic segment):

which is impossible, since

(rect. GD) = $\frac{2}{3}$ (parabolic segmt.).

Therefore

(portion of cyl.) is *not* less than $\frac{2}{3}$ (prism cut off).

But it was also proved that neither is it greater; therefore (portion of cyl.) = $\frac{2}{3}$ (prism cut off)
= $\frac{1}{6}$ (original prism).]

[**Proposition 15.**]

[This proposition, which is lost, would be the mechanical investigation of the second of the two special problems mentioned in the preface to the treatise, namely that of the cubature of the figure included between two cylinders, each of which is inscribed in one and the same cube so that its opposite bases are in two opposite faces of the cube and its surface touches the other four faces.

Zeuthen has shown how the mechanical method can be applied to this case[*].

In the accompanying figure $VWYX$ is a section of the cube by a plane (that of the paper) passing through the axis BD of one of the cylinders inscribed in the cube and parallel to two opposite faces.

The same plane gives the circle $ABCD$ as the section of the other inscribed cylinder with axis perpendicular to the

---

[*] Zeuthen in *Bibliotheca Mathematica* vii$_3$, 1906–7, pp. 356–7.

plane of the paper and extending on each side of the plane to a distance equal to the radius of the circle or half the side of the cube.

AC is the diameter of the circle which is perpendicular to *BD*.

Join *AB*, *AD* and produce them to meet the tangent at *C* to the circle in *E*, *F*.

Then $EC = CF = CA$.

Let *LG* be the tangent at *A*, and complete the rectangle *EFGL*.

H
L
V
A
X
G
M
N
O
Q
S
R
P
B
D
K
E
W
C
Y
F

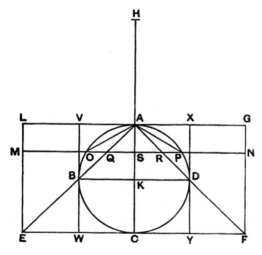

Draw straight lines from *A* to the four corners of the section in which the plane through *BD* perpendicular to *AK* cuts the cube. These straight lines, if produced, will meet the plane of the face of the cube opposite to *A* in four points forming the four corners of a square in that plane with sides equal to *EF* or double of the side of the cube, and we thus have a pyramid with *A* for vertex and the latter square for base.

Complete the prism (parallelepiped) with the same base and height as the pyramid.

Draw in the parallelogram *LF* any straight line *MN* parallel to *EF*, and through *MN* draw a plane at right angles to *AC*.

This plane cuts—

(1) the solid included by the two cylinders in a square with side equal to OP,

(2) the prism in a square with side equal to MN, and

(3) the pyramid in a square with side equal to QR.

Produce CA to H, making HA equal to AC, and imagine HC to be the bar of a balance.

Now, as in Prop. 2, since $MS = AC$, $QS = AS$,

$$\begin{aligned} MS \,.\, SQ &= CA \,.\, AS \\ &= AO^2 \\ &= OS^2 + SQ^2. \end{aligned}$$

Also

$$\begin{aligned} HA : AS &= CA : AS \\ &= MS : SQ \\ &= MS^2 : MS \,.\, SQ \\ &= MS^2 : (OS^2 + SQ^2), \text{ from above,} \\ &= MN^2 : (OP^2 + QR^2) \\ &= (\text{square, side } MN) : (\text{sq., side } OP + \text{sq., side } QR). \end{aligned}$$

Therefore the square with side equal to MN, in the place where it is, is in equilibrium about A with the squares with sides equal to OP, QR respectively placed with their centres of gravity at H.

Proceeding in the same way with the square sections produced by other planes perpendicular to AC, we finally prove that the prism, in the place where it is, is in equilibrium about A with the solid included by the two cylinders and the pyramid, both placed with their centres of gravity at H.

Now the centre of gravity of the prism is at K.

Therefore $HA : AK = (\text{prism}) : (\text{solid} + \text{pyramid})$

or $2 : 1 = (\text{prism}) : (\text{solid} + \tfrac{1}{3} \text{ prism})$.

Therefore $2\,(\text{solid}) + \tfrac{2}{3}\,(\text{prism}) = (\text{prism})$.

It follows that

$$\begin{aligned} (\text{solid included by cylinders}) &= \tfrac{1}{6}\,(\text{prism}) \\ &= \tfrac{2}{3}\,(\text{cube}). \quad \text{Q.E.D.} \end{aligned}$$

There is no doubt that Archimedes proceeded to, and completed, the rigorous geometrical proof by the method of exhaustion.

As observed by Prof. C. Juel (Zeuthen *l.c.*), the solid in the present proposition is made up of 8 pieces of cylinders of the type of that treated in the preceding proposition. As however the two propositions are separately stated, there is no doubt that Archimedes' proofs of them were distinct.

In this case AC would be divided into a very large number of equal parts and planes would be drawn through the points of division perpendicular to AC. These planes cut the solid, and also the cube VY, in square sections. Thus we can inscribe and circumscribe to the solid the requisite solid figures made up of element-prisms and differing by less than any assigned solid magnitude; the prisms have square bases and their heights are the small segments of AC. The element-prism in the inscribed and circumscribed figures which has the square equal to OP^2 for base corresponds to an element-prism in the cube which has for base a square with side equal to that of the cube; and as the ratio of the element-prisms is the ratio $OS^2 : BK^2$, we can use the same auxiliary parabola, and work out the proof in exactly the same way, as in Prop. 14.]

COSIMO is a specialty publisher of books and publications that inspire, inform, and engage readers. Our mission is to offer unique books to niche audiences around the world.

COSIMO BOOKS publishes books and publications for innovative authors, nonprofit organizations, and businesses. **COSIMO BOOKS** specializes in bringing books back into print, publishing new books quickly and effectively, and making these publications available to readers around the world.

COSIMO CLASSICS offers a collection of distinctive titles by the great authors and thinkers throughout the ages. At **COSIMO CLASSICS** timeless works find new life as affordable books, covering a variety of subjects including: Business, Economics, History, Personal Development, Philosophy, Religion & Spirituality, and much more!

COSIMO REPORTS publishes public reports that affect your world, from global trends to the economy, and from health to geopolitics.

FOR MORE INFORMATION CONTACT US AT
INFO@COSIMOBOOKS.COM

- if you are a book lover interested in our current catalog of books

- if you represent a bookstore, book club, or anyone else interested in special discounts for bulk purchases

- if you are an author who wants to get published

- if you represent an organization or business seeking to publish books and other publications for your members, donors, or customers.

COSIMO BOOKS ARE ALWAYS
AVAILABLE AT ONLINE BOOKSTORES

VISIT COSIMOBOOKS.COM
BE INSPIRED, BE INFORMED